I WANT TO BE A...

SCIENTIST

DOUG BRADLEY

PowerKiDS press.

New York

Published in 2023 by The Rosen Publishing Group, Inc.
29 East 21st Street, New York, NY 10010

First Edition

Editor: Caitie McAneney
Book Design: Rachel Rising

Photo Credits: Cover, p. 1 Pressmaster/Shutterstock.com; pp. 4, 6, 8, 10, 12 ,14, 16, 18, 20 april70/Shutterstock.com; p. 5 Maksim Shmeljov/Shutterstock.com; p. 7 Gorodenkoff/Shutterstock.com; p. 9 montira areepongthum/Shutterstock.com; p. 11 ESB Professional/Shutterstock.com; p. 13 Nicole Helgason/Shutterstock.com; p. 15 Evgeny Haritonov/Shutterstock.com; p. 17 wavebreakmedia/Shutterstock.com; p. 19 Frame Stock Footage/Shutterstock.com; p. 21 Twinsterphoto/Shutterstock.com.

Some of the images in this book illustrate individuals who are models. The depictions do not imply actual situations or events.

Library of Congress Cataloging-in-Publication Data
Names: Bradley, Doug.
Title: Scientist / Doug Bradley.
Description: New York : PowerKids Press, 2023. | Series: I want to be a... | Includes glossary and index.
Identifiers: ISBN 9781725339897 (pbk.) | ISBN 9781725339910 (library bound) | ISBN 9781725339903 (6pack) |
ISBN 9781725339927 (ebook)
Subjects: LCSH: Scientists--Juvenile literature. | Science--Vocational guidance--Juvenile literature.
Classification: LCC Q147.B656 2023 | DDC 509.2'2--dc23

Manufactured in the United States of America

CPSIA Compliance Information: Batch #CSPK23. For Further Information contact Rosen Publishing, New York, New York at 1-800-237-9932.

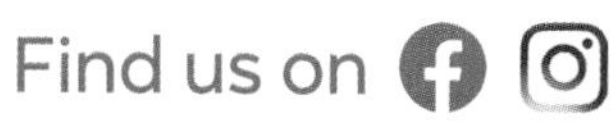

CONTENTS

What Do Scientists Do? 4
Where Do Scientists Work? 6
Chemists 8
Environmental Scientists 10
Biologists 12
Geologists 14
How to Be a Scientist 16
Making a Difference 18
Asking Questions! 20
Glossary 22
For More Information 23
Index 24

What Do Scientists Do?

Scientists work in all different areas. Some study space, weather, water, or animals. One thing scientists have in common is that they ask hard questions. Then they use **research** and **evidence** to guess the answer. They test that answer. Then they share what they've learned.

Where Do Scientists Work?

Scientists work in many different settings. Some work in a lab. That's a place with many science tools. They can test things in the lab. Others work in the field. That means they go out to gather or **observe** things, like rocks or water. Some do both.

Chemists

Chemists are scientists who research **chemicals**. They see what happens when you mix two or more things together. They use what they've learned to make things. Chemists can make anything from **medicine** to food to shampoo!

Environmental Scientists

Environmental scientists work to keep the environment, or natural world, safe. Some go out and gather samples, or small bits, of water or soil. Then, they test them in the lab. They see how things can harm the environment. Then they try to fix it.

Biologists

Biology is the study of life. Biologists study living things. They study people, plants, and animals. They see how certain things can help or harm other things. Some biologists work in labs and others work in the field. Marine biologists study living things in water.

Geologists

Geologists study Earth. They study the history of Earth and how it was made. They look at mountains and other kinds of land. They study soil and rocks. They can use what they know to keep people safe from natural events, such as **earthquakes** and floods.

How to Be a Scientist

Scientists study science in college, or school after high school. Some go to college for four years, while others go for much longer. They decide what area of science they want to work in. They take many classes about that kind of science.

Making a Difference

Scientists make a big difference in the world. Many scientists make things we use every day. Some make medicines that save lives. Some help farmers grow crops. Some track illnesses as they spread. Some track weather so people know what might happen.

Ask Questions!

Do you want to be a scientist? You can start by asking questions. When you want to know the answer, make a guess. Then, test out different answers. Share what you know with others. And always be open to learning more about something!

GLOSSARY

chemical: Matter that can be mixed with other matter to cause changes.

earthquake: A shaking of the earth.

evidence: Something that shows that something else is true.

medicine: A drug that a doctor gives you to help fight illness.

observe: To look at something closely.

research: Careful study to find new knowledge.

FOR MORE INFORMATION

BOOKS

Brundle, Joanna. *My Job in Science*. King's Lynn, England: BookLife Publishing, 2020.

Gaertner, Meg. *Great Careers in Science*. Lake Elmo, MN: Focus Readers, 2022.

WEBSITES

Kids Environment, Kids Health

kids.niehs.nih.gov

Discover the connection between the health of the environment and your own health with this fun resource.

Types of Scientists

easyscienceforkids.com/best-types-of-scientists-video-for-kids/

Learn more about different kinds of scientists, from botanists to mineralogists.

INDEX

A
animals, 4, 12

B
biologists, 12

C
chemists, 8
college, 16

E
earthquakes, 14
environment, 10
environmental scientists, 10
evidence, 4

F
field, 6, 12

G
geologists, 14

L
lab, 6, 10, 12

M
medicine, 8, 18

Q
questions, 4, 20

R
research, 4, 8

S
space, 4

T
testing, 10, 20
tools, 6

W
weather, 4, 18
water, 4, 6, 10, 12